Thorny Devil

by Grace Hansen

AUSTRALIAN ANIMALS

Abdo Kids

Abdo Kids Jumbo is an Imprint of Abdo Kids
abdobooks.com

abdobooks.com

Published by Abdo Kids, a division of ABDO, P.O. Box 398166, Minneapolis, Minnesota 55439.
Copyright © 2020 by Abdo Consulting Group, Inc. International copyrights reserved in all countries.
No part of this book may be reproduced in any form without written permission from the publisher.
Abdo Kids Jumbo™ is a trademark and logo of Abdo Kids.

Printed in China

052019

092019

THIS BOOK CONTAINS
RECYCLED MATERIALS

Photo Credits: Alamy, Animals Animals, Getty Images, iStock, Minden Pictures,
Shutterstock, ©Fritz Hiersche p.13/123RF.com

Production Contributors: Teddy Borth, Jennie Forsberg, Grace Hansen
Design Contributors: Dorothy Toth, Pakou Moua

Library of Congress Control Number: 2018963337
Publisher's Cataloging-in-Publication Data

Names: Hansen, Grace, author.

Title: Thorny devil / by Grace Hansen.

Description: Minneapolis, Minnesota : Abdo Kids, 2020 | Series: Australian
    animals | Includes online resources and index.

Identifiers: ISBN 9781532185472 (lib. bdg.) | ISBN 9781532186455 (ebook) |
    ISBN 9781532186943 (Read-to-me ebook)

Subjects: LCSH: Lizards--Juvenile literature. | Reptiles--Australia--Juvenile
    literature. | Animals--Australia--Juvenile literature.

Classification: DDC 597.955--dc23

Table of Contents

Thorny Devil

Thorny devils live in central
and western Australia.
They can be found in sandy
deserts and dry **scrublands**.

Thorny devils are lizards. And like many Australian animals, they are very **unique**. Thorny devils are almost entirely covered in pointy spines!

The pointy spines are for more than just looking cool. They help protect the lizard from danger.

The thorny devil's skin can change colors. This helps it hide.

Food & Hunting

Thorny devils spend most of their time looking for food. Their favorite food is black ants! They eat thousands of ants per day.

13

Their sticky tongues make

hunting and catching ants easy.

They have special teeth for

chewing the ants up.

Burrows

On very hot and cold days, thorny devils dig burrows. They find shelter inside and do not move much. This helps them survive.

Thorny Devil Babies

Females also dig burrows to lay their eggs. They lay 3 to 10 eggs at a time. After around 90 to 130 days, the eggs hatch.

Baby thorny devils are very small.
They eat their own egg shell for
strength. Then they climb out of
the burrow in search of ants.

More Facts

- Rain water collects between the lizard's spines. The water then moves toward its mouth. This is helpful in an almost waterless desert.

- Thorny devils can puff themselves up to appear bigger. This can stop predators from eating them because it makes them harder to swallow.

- These lizards walk in a funny way. They walk slowly, stop often, and rock back and forth. This may be another way they confuse predators.

Glossary

female – an animal that produces eggs or gives birth to live young.

scrubland – land covered with scrub.

unique – being the only of its type.

23

Index

Visit **abdokids.com**
to access crafts, games,
videos, and more!